The Little Book of UK Heists

Kitty Lyons

**PEBBLE POCKET BOOKS THREE
2021**

This work is licensed under a Creative Commons Attribution-NonCommercial-NoDerivatives 4.0 International License.

**ISBN 9781716328534
COBBLEBOOKS.WORDPRESS.COM
FACECRACK@RISEUP.NET**

CONTENTS

city slickers 01

small is better? 09

magpies and sparkly things 17

what a load of bullion 39

the grim 49

pretty pictures 59

the heist that broke the peace process 75

you either laugh or you cry 95

CHAPTER ONE

Often the the germ of an idea for a heist will come from an insider, from someone working at a jewellery shop or bank. Perhaps they are dissatisfied with their job, perhaps they are simply greedy, in any case they see a flaw in the security measures and suggest a cunning plan to their shadiest of friends. Of course we only hear about the ones that get caught, but it does introduce a weakness to the best of plans since the rozzers will be looking for a person on the inside right from the start of their investigations. The lesson of this book is that many insiders do get caught and big heists do often get busted. But not always.

In the case of the 1990 City bonds robbery in London, there might have been an insider passing on information, although if there was they were never publicly revealed, but what we do know for certain is that an earlier theft seeded the idea for the heist in the minds of some sophisticated criminals. The robbery that occurred later in the year was one of the largest ever on UK soil and the repercussions were global in scope. Some heists might be linked to Italian mobsters, Colombian narco-terrorists and the IRA, but few are linked to all three, as well as Dunstable Town football club!

The heist began with what appeared to be a mugging on 2 May 1990 in the City of London, the main financial district of the UK. As a man traipsed along a small street, his bag was ripped away from him and the perpetrator flashed a knife before sprinting away. Rather unusually for a simple mugging, the bag contained almost £300 million in bearer bonds. These bonds are certificates issued by financial institutions which entitle the person carrying them (ie the bearer) to the sum written on the paper, no questions asked. In this particular case, the bonds were being used as a part of an antiquated system which kept the UK financial system liquid. The courier was robbed of 301 bonds, of which £170 million were issued by the Exchequer and £121.9 million were bank certificates of deposit. Just in case anyone is starting to have any ideas about hopping on a bus to the City of London and trying their luck, this ludicrous system of financial balancing was discontinued in 2003 and the process is now totally digitised.

The crime syndicate which had ordered the mugging hoped it would look like an accident and that the Bank of England would drag its heels before cancelling the certificates, giving it time to fly the bonds to Zurich and cash them out. It had been inspired by a January 1990 newspaper story which reported that a courier had dropped £4 million in bearer bonds in the City of London, and they had then been handed in by a good samaritan. How did they know where and when the courier would be? Well that's an interesting question and suggests insider knowledge. In any case, the intricate plan to launder the monies hit an immediate snag when the Zurich connection had second thoughts, perhaps spooked by the sheer scale of the heist.

Thus the syndicate were forced to pursue other options, and this brought them into contact with some colourful characters. First up is Keith Cheeseman, a charismatic geezer who was something of an expert at petty fraud. In the 1970s, he achieved his childhood goal and bought Dunstable Town football club with his ill-gotten gains. And for a time the good times rolled, Cheeseman supplying the cash to sign legends George Best and Jeff Astle to the non-league club. Best never played competitively for Dunstable but as a publicity grabber it was priceless and Astle actually scored 34 goals. One Christmas, Cheeseman went to the manager and told him he was planning a festive bash for everyone at the club, so he needed the names and addresses of everyone to send out the invites. The party never happened and instead

Cheeseman took out loans in the name of every person. Unsurprisingly, he was eventually found out and in 1977 he was sent to prison for six years. Always the showman, he had arrived at court in a silver chauffeur-driven Rolls Royce, but the glory days of Dunstable FC were over and it went bankrupt. Upon his release, Cheeseman quickly returned to his old tricks, seducing a female financial services consultant and plotting with her to steal £8 million from her employers. They were busted and he went back to jail for another three years.

It's thus unsurprising that Cheeseman became involved with the scheme to launder the bonds. He contacted his old acquaintances Mark Lee Osborne and Raymond Ketteridge. Ketteridge quickly got to work only to be arrested on the charge of handling £70 million in stolen bonds. Somehow the charges were laid aside and he later avoided getting extradited to the US. Perhaps he turned informant? Osborne didn't fare much better, he met what he thought was a mob representative in a bar in New York, unfortunately for him the contact was an FBI informer. Osborne was arrested and quickly implicated Cheeseman, who was arrested at his home in London and bailed. When Osborne turned up dead in the boot of a car with two bullet holes in the back of his head, Cheeseman

City slickers

freaked out and skipped bail to Tenerife. In a bizarre side-story, he was for a time considered to be the Bolney torso, a still unexplained headless body discovered in a Sussex village by a man taking a pee on his way home from the pub. However, the corpse was not Cheeseman, since the FBI found him alive and well, sunning himself on a beach in Spain. He was extradited to the US and faced charges for laundering bonds from the City heist and also a big robbery in New York.

Patrick Thomas is the man thought to have been the mugger. He was a small-time south London criminal who was shot dead on his doorstep in December 1991. Police at first thought he was victim of a drug trade war, until they pieced together his connection to the heist. They still hadn't nobbled anyone directly involved in planning the robbery and in fact still haven't. The shadowy figures have stayed hidden and although the police claim to have eventually recovered all but two of the bonds, it's not known how much money was made by using them as collateral.

City slickers

CHAPTER TWO

The Securitas depot robbery is the largest cash robbery in UK history. The gang got away with over £53 million and actually left another £150 million behind because they didn't have space to take it with them! Events began when the manager of the depot was pinged by an unmarked police car with blue flashing lights as he drove home on a Tuesday evening. He pulled over (as you would do) and followed the fake cop into the other car (???), whereupon he was abducted. He was transferred to a van and tied up, then taken to a farm. Meanwhile two other gang members had kidnapped his wife and eight year-old kid from his home in Herne Bay and they also turned up at the farm. The manager was threatened at gunpoint and told to comply with their demands.

At 1am, the gang set off in a convoy of vehicles including a 7.5 ton truck and headed to the Securitas depot which was located at Vale Road, in Tonbridge. One gangster still dressed as a cop went inside with the manager and opened the gate for the vehicles. Once inside the gang of at least six people threatened the staff with weapons such as a Skorpion submachine gun and an AK47 (which most likely ensured their compliance). The fourteen members of staff and the family were all tied up then locked into cash storage cages.

Then in under an hour, the gang loaded up the truck with £53,116,760 in Bank of England notes!! They made good their escape leaving the staff locked up. Accounts differ on the how but the staff somehow managed to raise the alarm another hour later. When the police arrived they found the shaken but unharmed employees and CCTV footage of the criminals. The police announced the details of the heist, at first underestimating the amount stolen.

The heist had been meticulously planned, but the cops were helped by several very basic errors made during its execution. Forensics discovered DNA at the crime scene and the vehicles which had been used began popping up in pub carparks. One car even still had £1.3 million in it! The first arrests occurred already by the Thursday, with three people being lifted in London, one for trying to deposit money at a building society in Bromley which still had the Securitas wrappings on it!

On the following Saturday, the police raided two houses near Tunbridge Wells where they discovered plans of the depot. It seems a bit silly to leave them lying around after the heist has taken place. The next day, the two men whose houses had been raided were arrested after a chase when the police shot out the tyres of their car. The cops then hit a small jackpot in March when they raided a car yard in Welling and found a stash of £7 million. The owner of the yard was quick to announce that he was as gobsmacked as anyone else and would be happy to help the police with their enquiries since he was subletting it.

By June, over 30 people had been arrested, including Lee Murray and Paul Allen in Morocco.

Murray was a former cage-fighter alleged at court to have been the ringleader of the heist. Police sources claimed that it was him wearing a prosthetic disguise who had impersonated a copper to abduct the manager and also that he was the man seen on CCTV at the depot urging everyone to work fast and forcing them to leave after just one hour. Murray denied all involvement and fought his extradition from the UK, yet he had managed to incriminate himself ... he had left the UK for Morocco soon after the heist, but in the time in between he managed to crash his car and to leave his mobile phone in it when he fled the scene. This error was then exacerbated by the rather odd news that he had accidentally recorded a telephone call to an accomplice in which he incriminated himself?! Murray beat extradition after proving his dad was Moroccan but then got ten years in jail there, time enough to brood on how he managed to convict himself. Allen was extradited from Morocco to the UK and got eighteen years. Out after nine, he was required to pay back £1.2 million but only paid £420 and had his debt wiped on appeal. His luck caught up with him when he was shot several times at his home in north London in 2019, yet he survived and associates said they pitied the gunman for not succeeding in killing him. When the case came to trial at the Old Bailey in London (England's Central Criminal Court), in June 2007, eight people were charged with a range of offences. Michelle Hogg quickly decided she could not be arsed with jail and decided to shop her former comrades. Hogg was a makeup artist who had been roped in to do the prosthetic disguises and she won her freedom (plus a life in witness protection) by giving evidence for the prosecution. Thanks to Hogg, five men went down for long sentences. These were Jetmir Bucpapa, Roger Coutts, Emir Hysenaj, Stuart Royle and Lea Rusha.

Mr Hysenaj was the inside man! He was an Albanian (like Bucpapa) who had moved the UK after marrying an English woman and was working at the depot. He had filmed the inside of the building for the gang using a sophisticated spy camera in his belt (I want one). Ex-criminals interviewed by Vice on the ten year anniversary of the heist remarked that there had been too many people involved, pointing to Hogg as an unnecessary weak link. They praised the use of an inside man, although to me that seems like another obvious weak link. What was interesting was that they all poured scorn on the idea that big heists are good for business, saying that smaller heists are much less likely to get busted.

Most (but not all) the people did get caught in this case but we also have to remember that over £30 million has never been recovered and several men are thought to have disappeared, each carrying substantial chunks of the loot.

Small is better?

CHAPTER THREE

When one thinks about heists, car chases and diamonds quickly come to mind and I apologise for the lack of both so far in this book. As we turn our eyes to London for this chapter, there'll be a lot of jewels and fast cars. Historically, the areas of both Hatton Garden in Holborn and Bond Street in Mayfair have hosted many jewellery shops and workshops. And it is not surprising at all that enterprises dealing with extremely expensive items would attract criminals.

One high-end jewellers which has had more than its share of robberies is Graff, started by Laurence Graff in 1960. In 1980, two Chicago gangsters robbed its Sloane Street store of £1.5 million worth of gems, including the Marlborough diamond, which was never recovered. Joseph Scalise and Arthur Rachel escaped by taxi and quickly went to the airport, pausing only to mail a package, which is believed to have contained the Marlborough diamond. Various factors, such as Rachel renting the getaway vehicle in his own name and Scalise being positively identified at the store by the lack of four fingers on his left hand, meant that they were arrested as they stepped off their flight back to Chicago. By 1983, Scalise had been extradited back to the UK and the following year he was imprisoned for 16 years. After his release Scalise was arrested a second time with Rachel and another associate in 2010 for planning a series of heists; Scalise had been secretly recorded by the FBI boasting that no cops would ever recognise him, but this assumption proved wrong and he was jailed in 2012 for 106 months.

Over a decade later, Graff was again the victim of a sophisticated heist. This time it was the workshop in Hatton Garden which was attacked by the Rascal Gang, who made off with £7 million (they got the nickname because their getaway vehicle of choice was the Bedford Rascal van). The Graff store at New Bond Street was hit in 2003, to the tune of £23 million. This time around, suspicion quickly alighted on the Pink Panther gang. Named after the 1963 Peter Sellers comedy classic, the Pink Panthers were suspected to be Balkan ex-soldiers who had turned their considerable expertise towards a lucrative new industry. Their exploits have spanned the globe, and are recognised for their skill and cunning. Analysts have estimated the gang could be as large as 800 people, which is known in my circle of expert commentators as a big fucking gang. The people are connected in the same way as extremist groups in small autonomous cells to avoid spilling the beans on other members who they don't know and have never met. So who is at the top of the tree? We just don't know.

The New Bond Street heist hit the headlines owing to the size of the haul and actually was the very one which led to the nickname being bestowed on the group by Interpol officials, when it was discovered that a stolen diamond ring had been hidden in a jar of cream, just like in the 'Pink Panther'! Two smartly dressed men entered the store wearing disguises which staff initially explained away as them being celebrities not wanting to be recognised. Suddenly the men pulled out handguns and looted the store. Nebojsa Denic and Milan Jovetic were later arrested, but 80% of the jewels have not been recovered.

The Pink Panthers struck again at the Sloane Street store in 2005, when three men got away with £2 million, and the same store was again robbed of £10 million in 2007.

The Panthers were back again after another two year gap when a massive heist occurred in 2009. Two men robbed the New Bond Street Graff store to the tune of £40 million. The raid was meticulously planned, with the first step being to pay a professional make-up artist to change their features, for which they said they were making a music video (ironically it was the same firm which had been tricked into doing make-up for the Securitas heist gang). At the store, the two men stole 43 items as if to order and made their escape, firing a gun as they left to sow confusion and chaos. They jumped into a BMW and quickly switched to a Mercedes, before changing again to a third car and getting away. However, the police had a lucky break when they found a pay-as-you-go mobile phone carelessly discarded in one of the cars. Tracing the calls, they were able to track down ringleader Aman Kassaye and several other gang members, arresting in total ten people. Kassaye was jailed for 23 years in 2010 and three accomplices got 16 years each. Nothing from the jewellery has so far been recovered.

Magpies and sparkly things

The Pink Panthers were also connected to the theft of the Portland Tiara in 2018 from the Harley Gallery in Nottinghamshire. This tiara was made for the Duchess of Portland to wear at the 1902 coronation of Edward VII and Queen Alexandra; it had never been valued, being described as priceless. The wisdom of the wise is that it will have been immediately broken up into its component parts and therefore will never be seen again. This was suspected to be yet another Pink Panther robbery since it bore one of the hallmarks, a burnt out Audi S5.

Whilst Graff has borne the brunt of the major heists, other jewellers have not escaped notice completely. The Bond Street branch of Chatila was robbed in 2010, leaving a bill of £1 million. The Boodles store, also on Bond Street, fell prey to an elaborate scam when a group of purported Russian businessmen met the Boodles chairman in Monaco and requested that he sell them seven diamonds. The deal was set and ready to go, subject to checks by a gemologist. So a meeting was set up in London, and a gemologist duly arrived. She was met by both the head of the store and the Boodles gemologist who set out the diamonds for examination. After the valuation had taken place, the diamonds were placed in a locked bag to be held by Boodles until payment was received, but when staff became suspicious after two days and opened the bag, they found seven small pebbles inside. They had been fooled! Of all the heists in the book, this is definitely one of the smoothest.

A more brutal heist occurred in 2019, when four men surprised an employee of Le Vian diamond company in a car park in Staines, Surrey. They knocked him to the ground and stole a briefcase containing over £4 million in jewels, including rare chocolate diamonds. All four were quickly apprehended after trying to sell the gems locally, which seems idiotic. They weren't completely stupid however; it emerged they had decided to follow the man after reading an announcement in the newspapers that the diamonds would be delivered in Staines.

Moving on to thefts from private residences, Tamara Ecclestone (the daughter of Bernie Ecclestone the Formula1 magnate and his wife Slavica Radić) had an estimated £50 million worth of jewels stolen from her house in Kensington in 2020. The mind boggles as to how one person could have so much jewellery lying around!

The notorious Johnson family gang have been linked to a chain of robberies of stately homes in the early 2000s. In 2003, they stole items including antique snuffboxes from the National Trust's Waddesdon Manor which came to a total of £5 million. Two years later, they robbed the house of another wealthy Formula1 businessman, namely Paddy McNally. He owned Warneford Place near Swindon, which 'James Bond' author Ian Fleming had renovated in the 1960s. The Johnsons also robbed Ramsbury Manor in Wiltshire in 2006, getting away with an estimated £30 million in various goods. At the time, it was the home of property baron Harry Hyams and the sheer scale of this heist prompted police forces to pool resources in order to catch and convict some members of the syndicate. Aside from stately homes they had specialised in ram-raid robberies of cash points and shops; after the convictions the figures for these sorts of crimes dropped dramatically.

Lest you might think heists are a recent phenomenon, in my researches I discovered that Hatton Garden has been the site of robberies just about as long as it's been wealthy. The post office was robbed of an estimated £80,000 in 1881, which in today's money would be a cool £8 million. It was an ingenious heist; at the end of a busy day in the post office, the lights suddenly went out, causing confusion until the gas connection was turned back on again. Soon afterwards, the mail clerks noticed that a bag of packages ready to be sent by registered post had vanished. The goods included diamonds, sapphires, emeralds and rubies! Since the amount stolen could only be estimated, it might actually have been a lot more.

The size of the haul can also only be guessed at in the Hatton Garden safety deposit burglary of 2015, since there was a vault which was used by local jewellers to store goods and was also a convenient dropping point for any gangsters looking for a supposedly secure place to keep their ill-gotten gains. The heist amused the tabloids no end, since first of all nobody could work out who had done it (no Pink Panther hallmarks to be seen) and then when the gang was arrested, it turned out to be a bunch of pensioners! But these were no ordinary grey-haired old biddies, these were seasoned gangsters doing one fateful last job. The Great British public seemed to be generally on their side since nobody was harmed and they only got caught through bumbling incompetence. What also caught the attention was that the last known member of the gang evaded capture for three years longer than everyone else. This was the mysterious Basil, of which there is quite a lot more to be said.

The idea for the safety deposit heist probably first hit ringleader Brian Reader (born 1939) when he was serving time in jail for his part in processing the loot from the Brinks-Mat robbery. Reader was an old school south London criminal who had developed a reputation for being able to cut his way into safes and vaults. Things went well enough for him to be able to buy a gold Rolls-Royce in the 1980s. After the 1983 Brinks-Mat heist, Reader and his associate Kenneth Noye were contacted to process the stolen gold since the two already had a bullion VAT scam going on. They were happy to be involved, not knowing the problems ahead. Noye was notorious for his temper and in 1985, was arrested for stabbing an undercover cop he discovered in his garden, when Reader was present. At the ensuing murder trial, the jury found Noye had acted in self-defence and thus Reader escaped conviction as an accessory. However, a year later the two men were convicted of charges relating to their gold processing activities and Reader received nine years. So he had plenty of time to think about his next heist. Noye got fourteen years and after serving his time was subsequently jailed for life in 2000 for a road rage murder on the M25 (he got released in 2019, life not being life any more in the UK judicial system).

The vault at 88-90 Hatton Garden was infamous for being a place where vast amounts of expensive objects were stored. It was common knowledge that local jewellers had boxes where they kept items they were working on or that were too valuable to keep in their shop, and it was less well-known that criminals used the vault as well to store items. Over the years, Reader came to learn that a major London crime syndicate had a special interest in the vault. This was bad news since it meant that if he robbed the vault without asking their permission, he would no doubt be executed since the Adams family did not mess around. And Reader had only to think of 'the curse of Brink's-Mat' in which several people linked to the heist have ended up dead. The problem was that if he shared his plan with the syndicate, what was to stop them simply taking it over? Reader wrestled with these questions for years, patiently keeping his eye on the vault all the time. He talked them over with his old associate Terry Perkins (born 1949).

In 2003, the vault was in the news for a startling individual crime. A man known as Philip Goldberg had taken a deposit box and was a frequent visitor after taking jewels to be valued at nearby shops. His last recorded visit was a Saturday in June 2003 and on the following Monday it was discovered he had stolen at least £1.5 million. Reader must have read all the newspaper reports with interest and one thing must have stuck out - there was no CCTV in the vault! Another spinoff from the burglary was that Perkins and Reader were summoned to a meeting with the crime syndicate bosses. The Adams family had heard on the grapevine about their daydreams and made it crystal clear that the two men were not to touch the vault without their explicit permission since they held items there.

Despite all of that, by 2013, Perkins and Reader

had decided it was time for one last heist. They were both not getting any younger and were both in need of cash. They agreed to talk to the syndicate. Therefore in late 2013, they flew out to Spain to talk to a representative. He proposed that they rob the vault with the syndicate's permission and that they only had to one thing in return, which was to steal one specific box and make sure the family received it untouched.

He then introduced them to Basil. The character of Basil was for a long time a mystery, since when the group was eventually arrested, he was not. The rest of the group's silence is now more understandable, since Basil was under the protection of the syndicate. We'll have more on Basil later, suffice it to say for now that Basil guaranteed the gang access to the vault since he had the keys to the front door. In 2013, there had been a break-in at home of the son of the owner of the vault. The Hatton Garden Safety Deposit Company was owned by a Sudanese family and the son was the day-to-day manager. It's never been officially established if there was a link, but it seems likely the syndicate robbed the house for the keys.

Perkins and Reader now needed a gang, so they rustled one up. It included Danny Jones, Kenny

Collins, Carl Wood and Billy "The Fish" Lincoln. The youngest man out of all of them was a spritely 48! They set about robbing construction sites for the tools they needed, including a specific drill which can cut through a 2 metre thick wall of concrete. In early 2015, they met Basil again and he gave them detailed schematics of the building. Then he stunned them by telling them that the syndicate only wanted them to take certain boxes. This was a new condition and non-negotiable. The gang did not like this at all, but it was too late to do much about it and the heist was already planned for the Easter bank holiday.

Now after all the many years of daydreams and careful planning, after all the discussions and worries about the syndicate, the game was finally on! Basil entered the bank first and cut the wires on the alarm. He let the rest of the gang in through a fire exit and they scrambled down through the disabled lift shaft into a room next to the vault. The drill took three people to carry it! They got to work, putting the drill against the 2 metre thick wall of the vault. Meanwhile, unbeknownst to the them, the alarm had actually been tripped, since the cutting of the wires had triggered an security measure which automatically called the police. The cops ignored the robotic message for three hours then simply notified the security company, which sent a guard round. The lookout warned the men inside to stop drilling whilst the guard lazily checked the outside of the building. He assumed there had been an electrical fault with the alarm and in a crucial mistake told the manager who was five minutes' away that it was all good and he could go home instead of checking inside.

Not knowing just how close they had come to getting busted, the men carried on working. By 5am, they were through the concrete, only to be met (as expected) by a sheet of steel on the inside of the vault. The tool to blow this off the wall was a hydraulic ram, which must have been crazy loud in the small room. When the ram failed there was consternation, until Basil spoke up and suggested they leave and come back again the next night with more equipment in order to finish the job. This had not been part of the plan and it must have felt like madness to leave a crime scene only to return to it again 24 hours later, but the gang had little choice. They were all committed to the heist and further, all of them knew that their lives were at risk if they didn't do what the syndicate wanted. So they decided to return the following evening, but when the 24 hours had passed, there were only four left out of six. Reader the leader had decided his health was simply too bad to risk another night of exertion and another member was too scared to carry on. That left four, with the new tools plus Basil. They got to work again and finally they were in! Basil and Danny Jones squeezed in through the hole made by the drill and it was like Christmas as Jones started smashing open boxes. Basil on the other had, went straight to the specific box he wanted and put it in a bag. By 5.45am the gang had loaded up the van with the loot and headed off.

So were they ecstatic to have got away with it? Not in this case. Reader of course was not even there and fretting as to whether he would ever see his share. Perkins assured him he would, but there was always the danger that the syndicate might double cross them and take everything. The gang were busy processing the loot when Perkins got the call he was dreading. The syndicate told him there had been a change of plan and that Basil would be passing by to collect the most valuable jewels for "safekeeping". There was nothing they could do when Basil arrived a short time later accompanied by two heavies who made no attempt to hide their guns. Despite assurances that they would see the money later for the jewels, it was obvious that the gang had been diddled and equally obvious that there was nothing they could do about it. In fact there was very little preventing the syndicate from grassing them up or even eliminating them now their usefulness was exhausted.

When news broke on the Tuesday after Easter about the heist, the newspapers were all over it. The police were initially bamboozled and were forced to admit an alarm call had gone unanswered. The rumour mill went wild! The police stated tracking the owners of the safety deposit boxes, both stolen and still present, yet found owners oddly reticent to come forward and quickly realised there was a criminal angle to the vault. One week after the heist, the 'Daily Mail' scooped the police by releasing CCTV footage that everyone else had missed. This showed the gang entering and leaving. Just like in the movies, they were all given nicknames: the Gent, the Tall Man, the Old Man, the Ginger Man, Mr Strong and Mr Montana.

Now everybody knows what the gang should have done was to keep their heads down, but of course the circumstances were weird, since firstly Reader had not been present on the second night and naturally wanted to know what had happened, and secondly they were all scared of the syndicate coming after them. Would 'the curse of Hatton Garden' become a thing? Done differently, the heist might have remained unsolved, but the ageing gang made some fundamental errors. They had wisely put their mobile phones aside before the heist, but afterwards they started to use them again and called each other without a thought. Danny Jones had hired the extra equipment with his own bankcard and the gang had left pretty much all their tools behind in the vault. As the tabloid frenzy continued, the men could not resist meeting up and whilst the police searched far and wide for suspects, the criminal histories of Reader and Perkins brought them under suspicion. There were just too many similarities to previous heists such as the Securitas Depot and Baker Street for this not to be an avenue of exploration. One retired cop even thought he recognised Perkins on the CCTV footage.

Soon the careful police work got results, with Kenny Collins' Mercedes being flagged up as a frequent visitor to Hatton Garden in the weeks previous to the heist. Collins' criminal record was enough to arouse suspicions and surveillance on Collins showed him to be meeting with none other than Perkins and

Reader on a regular basis. The police were also hampered by their own prejudice, believing that old school criminals like Perkins and Reader would not collaborate on a project, being gangsters from different side of the Thames. And of course they were assumed to be too old to be squeezing through holes in vaults. Yet when the cops taped the conversations of the old lags in the pub, they were staggered to realise that they were not only talking about the heist, they were crowing about actually having done it, happily reminiscing over a pint!

On 19 May, 2015, the police swooped in a massive operation, arresting everyone at once at their own homes (except of course for Basil who was to remain at large until 2018). The Flying Squad knew they had to crack the case. When the arrests were announced and the tabloids realised how old the gang was (with a combined age of 533), they went mad. The men standing trial were Brian Reader (76), Terry Perkins (67), Kenny Collins (74), Jon Harbison (42), Billy Lincoln (59), Paul Reader (50), Carl Wood (58), Hugh Doyle (48) and Danny Jones (58).

Harbison was quickly released having proven he was just a taxi driver. For Paul Reader his arrest must have come as a bit of a surprise as well. Since his father didn't trust mobile phones and never had one, the son had become the person to call if the rest of the gang wanted to reach his dad. And for that he was remanded for months into pre-trial detention. Whilst at Belmarsh, the main members of the gang were contacted by the crime syndicate "thanking" them for staying silent, so there was no question of implicating them or Basil, since the potential repercussions were obvious.

Reader got six years and was released in 2018 at the age of 79, owing to ill health. Danny Jones was jailed for seven years and when he failed to comply with a later order to repay his ill-gotten gains he was given another seven. Likewise for Kenny Collins and Terry Perkins. Soon after the repayment order was made, Terry Perkins died of a heart attack in Belmarsh Prison at the age of 69.

Basil got ten years after his arrest in 2018. It emerged at his trial that Basil (real name Michael Seed) was in his fifties and lived in Islington. After a drugs conviction in the 1980s, he had dropped out of view completely, living in a council flat, paying no tax and taking no benefits. When his flat was raided, it was reported to be full of gadgets such as frequency jammers and he was busy breaking down jewellery which had been stolen from the vault. Whilst he always denied he was Basil and claimed not to know where the items had come from, he was identified on the CCTV footage by a gait specialist and was convicted. In 2020, Basil was given his own confiscation order in which he was required to repay almost £6 million within three months or face another seven years being added to his sentence.

The crime syndicate were not implicated by anyone. Only rumours suggest that the box stolen to order by Basil contained evidence linking his bosses to gangland murders.

What a tale! It will not be a shock dear reader to discover that there have already been three movies telling this bizarre tale: 'Hatton Garden: The Heist' (2016); 'The Hatton Garden Job' (2017), and 'King of Thieves' (2018). There have also been a TV show, radio play and several books. And now a little book as well!

Magpies and sparkly things

CHAPTER FOUR

As we saw in the previous chapter about sparkly things, people have always found it hard to resist nicking gold. Personally I'm not sure what the attraction is with this shiny metal and I would have no idea how to process it. And that's why I leave bullion heists to other people who are both bolder and better connected than me.

Between the world wars, Croydon Airport served as London's main airport hub. We are used to Heathrow and Gatwick ruling the roost, but they only came to prominence because Croydon was unable to expand! In the 1930s, Imperial Airways (which would later become British Airways following a series of mergers) served the British Empire, transporting mail, freight and gold bullion around the world. At night, the goods were kept under lock and key, guarded by a lonely watchman.

On 6 March 1935, the watchman discovered at 7am that the airport had been robbed! Between his last check at 4am and then, a gang had entered the airport and plundered the vault. They probably would have got away with it, had a curious early morning cyclist not noted down the numberplate of their getaway taxi and supplied it to the police. This led to the arrest George Mason, the taxi driver, who pleaded ignorance of the plot. He said a man he knew only as Little Harry had booked the cab and gave the cops several other names, namely Cecil Swanland, Silvio Mazzarda and John O'Brien. These men were duly arrested and at trial the prosecution made the case that Mason had driven the men to Croydon where they loaded the cases of gold bullion into the taxi and that he had then taken them to Swanland's house, where the landlady helpfully confirmed that they had unloaded items from the taxi. To make matters worse for Swanland, the police found seals in his apartment that had been used to secure the bullion and he was unable to explain a recent splurge on cufflinks and clothes when he was officially unemployed.

Swanland went down for seven years, but Mazzarda and O'Brien walked free because Mason first said he recognised their faces and then changed his story to say he had never seen them before. It seems likely that Mason was nobbled, since Mazzarda was connected to the notorious Sabini gang. The police reinterviewed Mazzarda two years later and secure in his protection, he told them that they had used an inside man at the airport who had given him a full set of keys. Surprisingly in the light of that admission, there were no more convictions.

What's more, none of the gold was ever found and Little Harry never faced justice. The gang had successfully stolen over £20,000 in bullion, estimated at around £1.2 million at today's prices, with their only costs being the payoffs to police, the inside man and Mason. Not forgetting of course the cost of the taxi journey to and from Croydon.

The Great Gold Robbery of 1855 was perhaps the first modern heist. A South Eastern Railways (SER) train was transporting gold bullion from London to Paris. It left London Bridge station and headed to Folkestone, whereupon the goods were transferred onto the railway steamer to Boulogne, before being put back onto another train heading to Gare du Nord in gay Paris. It was a regular service, running four times a day and when bullion was being transported it was always on the last train of the day, which left London in the evening.

Career criminal Edward Agar and former SER worker William Pierce had been discussing stealing the bullion for some time. With Agar's confidence and Pierce's insider connections, in 1855 they decided the bullion was ripe for the picking. Agar stipulated that he needed the keys for the safe and said he could handle all the other arrangements. Pierce brought in two men who were working for SER, namely William Tester and James Burgess. With Tester's help and ingenuity, Agar got hold of the facsimiles of the keys and then travelled on the train several times when Burgess was working as the guard, so he could make sure the keys worked in the safe.

Over months, the plan was honed to perfection, the gang even taking lead shot from the Tower of London to put in the bullion cases so that they would weigh the same amount as when originally packed and the theft would take longer to discover. All the men had to do now was to wait until a train was transporting bullion. On 15 May 1855, Burgess went for a smoke at the entrance of London Bridge station and wiped his face with a large white handkerchief, the signal for Agar and Pierce that today was the day. They immediately bought tickets and boarded the train. When it pulled off, Agar sloped off and opened up the safe. He went to work on one of the bullion cases, prying it open and taking out the gold bars. Using the scales he had brought with him, he replaced the same weight in lead shot and closed the container again with an improvised seal.

At the first stop at Redhill, Tester left the train carrying the stolen gold and went back up to London on another train. When the train pulled off again, Agar went back to work on two more cases. One was full of American Eagles ($10 gold coins) and in the other were more bars. He again took the loot and replaced it with shot, but hit a problem, because the second box of bars was too light, so he was forced to put some bars back. Helped by Pierce, he then put the cases back in the safe and cleared up the mess they had made. They then hid the gold in bags and went back to the first class travelling compartment they were supposed to be travelling in. The continued by train from Folkestone to Dover where they disembarked, taking their bags with them. They enjoyed a supper on the seaside in Dover and went for a walk, allowing Agar to throw his keys and tools into the sea, before taking the 2am train back up to London.

The crime was discovered in Paris the same evening when the cases were weighed and there were discrepancies with the original weight recordings. The cases were then opened to reveal the lead shot. The authorities in France were baffled by the heist, saying it was impossible and it must have occurred in England. Likewise, the English authorities said there was no way it could have happened before the border. The staff present that night were scrutinised to see if it was an inside job and Tester benefitted from having been seen back in London before the train had even reached Folkestone. Burgess was regarded as above suspicion since he had worked for SER for fourteen years.

An award of £300 was announced for information leading to the arrest of the thieves, who in the meantime were keeping themselves busy processing the loot. The Eagles were quickly sold and Pierce and Agar took to melting down the bars into smaller units at Agar's house.

Agar had separated from his partner and mother of his child Fanny Kay and began a new relationship with a young sex worker. To smooth things over with her pimp, Agar lent him £235, but when he went to collect the loan, he was arrested for fraud. This led to him being convicted to penal transportation for life in September 1855. How frustrating it must have been to be falsely busted for fraud when he was sitting on a fortune in gold! Whilst in prison awaiting transportation, Agar asked his solicitor to take all the money he had in his account (£3,000) and to give it to Pierce with the request for him to look after Fanny Kay and the child.

Pierce took the money, but did not give any to Kay, who was desperately poor. When it transpired that she had been ripped off, Kay went to the authorities and said she could tell them who had committed the Great Gold Robbery. When Agar's house was checked, gold flakes were found in the fireplaces, so he story was corroborated. Having found out that he had been betrayed by Pierce, Agar turned Queen's Evidence (not having anything to lose) and Pierce and Burgess were quickly arrested. Tester had moved to Sweden for a new job, so it took longer to arrest him.

The trial took place in January 1857 and was a media sensation. The three accused pled not guilty but with Agar's evidence they were always going to be convicted. John Chubb of Chubb locks gave evidence! After conferring for just ten minutes, the jury decided to convict all of the men. Burgess and Tester were handed down transportation to Australia for 14 years, and Pierce got a lower sentence because he was not an SER employee, which seems rather fortunate since he had previously been fired by SER for gambling! He got two years hard labour in England of which three months were to be spent in solitary confinement.

More recent bullion heists have included the KLM robbery of 1954, when £40,500 was taken and the A13 robbery in 1980, when a truck containing silver bullion was tricked into stopping and then ransacked to the value of £3.4 million!

Then of course there's the Brink's-Mat robbery, mentioned in the next chapter, in which the gang unexpectedly came across three tons of gold … what a load of bullion!

CHAPTER FIVE

It's all too easy when reading the latest tome in the esteemed range of pebble pocket books, reclining in my favourite armchair next to the log-fire, with a whisky in one hand and a cigar in the other, a light jazz standard tinkling away in the background, to forget that some heists do involve loss of life.

For me, the best heists are audacious and harm nobody. Even some robbers feel that way, for example Adam Worth and Brian Reader always wanted no violence (allegedly), but things can and do go wrong.

One notorious example is the Linwood bank robbery of 1969. Howard Wilson had been a cop, but when he failed to move up through the ranks as quickly as he believed a man of his fine brains deserved, he quit and went into business. He took on several shops in Glasgow and ran them into the ground, showing off that big brain power. By 1969, he was approaching bankruptcy, so he persuaded two drinking buddies, John Sim (an ex-prison warder) and Ian Donaldson (another ex-cop), that the best thing to do to solve their problems would be to rob a bank. They found a fourth man to be the getaway driver and did over the British Linen Bank in William Wood. They were in and out very fast, getting away with over £20,000, which on today's prices would be around £350,000.

However, the three friends had debts and six months later, they decided to it again. They had a problem this time around, because the young getaway driver wanted nothing to do with it. They went to see him on 23 December 1969 and that was the last time Archibald McGeachie was ever seen alive. It's always been rumoured that he ended up in one of the massive concrete pillars for the Kingston bridge across the River Clyde which was being constructed at that time.

One week later, the boys pulled off another heist. They raided the Clydesdale Bank at Renfrew, making off with over £14,000 (£250,000 today) and a metal box full of coins. They returned to Wilson's flat in Govanhill and took their loot inside. This is when their luck ran out, since an old police acquaintance of Wilson happened to be out shopping. Inspector Andrew Hyslop had never liked Wilson when he knew him on the force and when he noticed the three men acting suspiciously around the car, he decided to investigate. Luckily for him he called for backup first.

When Hyslop greeted Wilson, the gangster was affable and offered him a drink, after all it was the day before Hogmanay. Hyslop asked what was in the bags Wilson had been bringing in to the house and was told to go ahead and have a look. Hyslop only had a moment to register that he was looking at a huge pile of banknotes before Wilson stuck a gun on his temple and pulled the trigger. The gun jammed otherwise he would certainly have died, then Wilson reloaded and shot Hyslop in the head from a distance. The shots brought three more cops running, and Wilson, a trained marksman, shot two of them in the head as well. The third of the three hid in the bathroom whilst Wilson's accomplices Donaldson and Sim fled the scene in horror. Howard Wilson then stood over Acting Detective Constable Angus MacKenzie and delivered the coup de grace. He stepped towards Hyslop, preparing to finish him off as well, but then the cavalry arrived and he was disarmed. Hyslop somehow survived although he was confined to a wheelchair for the rest of his life. MacKenzie was dead and another cop later died of his injuries. Wilson admitted his guilt in court (becoming Scotland's first ever double murderer, legally speaking) and was put away for life. In jail, he was a feared hardman, before turning to literature and writing a crime novel. He was eventually released after almost 33 years inside in 2002. His release was controversial, with the relatives of his victims saying that they had a life sentence so they wanted one for him as well, without specifying who or what was being helped by keeping an old man in jail. The cops were all awarded medals for bravery.

In 1952, the Eastcastle Street robbery was a daring heist pulled off in daylight in central London, when a post office van was forced to stop and robbed of around £287,000 (about £8 million on today's prices). Just off Oxford Street, two cars shepherded the van to a stop and a gang quickly grabbed 18 mailbags and zoomed off. The heist captured the public imagination and nobody was ever caught, although in his memoirs organised crime-boss Billy Hill owned up to it. However, it should not be forgotten that the three post office workers in the van were all coshed and must have had traumatic memories of that day.

The grim

Another famed heist, probably the most famous one of all, is the Great Train Robbery of 1963 which netted £2.6 million (£100 million today) and made celebrities of the gangsters. Yet it should never be forgotten that they injured the train driver very badly. Most of the gang were caught and held to justice, although it is believed that at least another three people were never apprehended.

In the early hours of Thursday 8 August 1963, a travelling post office (TPO) train was nearing London from Glasgow. It was so called because a staff of 75 workers were sorting mail as they travelled through the night. When the train stopped for an unexpected red light in Buckinghamshire (caused by the thieves tampering with the signal), the three men driving the train were attacked and tied up, and the "high values packages" (HPV) wagon was stormed and robbed.

The gang then headed off to a derelict farm nearby, which had been bought for the purpose of sorting the loot, which came mainly as banknotes. As they had planned, the haul was larger than it usually would have been because of the bank holiday earlier in the week. They divvied it up, each person getting around

£150,000 (£3 million today), then listened to the news. When it seemed the police would soon be searching the area, the men agreed to leave the farm earlier than planned on the Friday rather than Sunday. The man paid to burn down the farm failed to do so and this handed the police a crime scene which yielded forensic results.

The gang had been planning the heist for months in exquisite detail but had been forced to bring more people on board, so from a starting group of about four there ended up being about fifteen people on the night. There are a whole host of books and other publications about both the heist and the individual characters, so I'll leave it to you to find out what happened. The heist is mentioned here because of the violence used against the train staff.

Early in the morning of 26 November 1983, an armed gang strolled into the Brink's-Mat depot near Heathrow Airport. They turned off the security system, then beat up the security guards and doused them in petrol, threatening to set them on fire unless they opened doors, which must have been horrible for them. The gang had been expecting to find cash, but they ended up with cash, diamonds and three tons of gold. The heist took £26 million in total, £100 million on 2020 prices.

The police quickly took to the inside job hypothesis because of how easy it was for the gang to gain access and they were right, Anthony Black was interrogated and confessed, then he implicated his former friend Mickey McAvoy. McAvoy had not hidden his newly acquired wealth, buying a mansion and naming his two Rottweilers Brinks ... and Mat. At trial McAvoy was given 25 years, as was his accomplice Brian Robinson, whilst another man walked free. This left the rest of the gang not accounted for, since at least six people had taken part in the raid. Those people have never been caught and most of the gold has never been recovered. Austrian police did arrest five men in December 1983 who they suspected were selling off gold bars from the heist, but in a bizarre twist, they discovered that despite having legit numbers printed on them, those bars were fake and made of tungsten!

The gold was perhaps as much of a burden as a good thing since the 'curse of Brink's-Mat' refers to remarkable number of deaths of people associated with the heist and its aftermath (most of whom have died after McAvoy came out of prison). The murders caused disquiet in the underworld since in at least two cases the victims showed no signs of suspecting they were under threat. Those who live by the sword die by the sword.

This chapter is dedicated to all the innocent victims of heists.

CHAPTER SIX

Over the years, many paintings have been stolen from UK galleries and private residences. Most recently, in March 2020, the Christ Church Picture Gallery in Oxford was robbed of three paintings. The most valuable of them was Anthony van Dyck's 'A Soldier on Horseback', the other two were 'A Boy Drinking' by Annibale Carracci and 'A Rocky Coast, with Soldiers Studying a Plan' by Salvator Rosa. The Christ Church gallery has around 300 paintings by the Old Masters and these three pictures alone are valued at around £10 million. Police suspected that the thieves made off by boat!

Rembrandt's 'Portrait of Jacob de Gheyn III' has the dubious honour of being the most stolen picture in the world. The diminutive 30cm by 20cm artwork was painted in the 17th century as one of a pair and now hangs in Dulwich Art Gallery in south London. It has been taken no less than FOUR times from the museum, an achievement worthy of the Guinness Book of Records. Firstly in 1981 it was robbed but didn't get far, the police intercepting the thieves in a taxi nearby. Second time round in 1983 it was gone for four years, turning up in a train station in Münster, Germany. Thirdly, it was discovered in a graveyard in nearby Streatham and the fourth and final time, it was retrieved from a cyclist furiously pedalling their escape. And thus the tiny portrait entered the record books.

The 'Portrait of the Duke of Wellington' was painted by Goya in the early 19th century. It depicts the leader after his victories in the Napoleonic Wars. It was auctioned off in the 1960s and nearly left the UK before being bought for the nation and deposited in the National Gallery in London. Just days later it was stolen. The police were mystified and it became a celebrated case, even popping up as a joke reference in the baddie's lair in the James Bond film 'Dr. No'. The plot thickened when the thief released a series of odd notes.

Firstly, a ransom of £140,000 was demanded. This was precisely the amount for which the painting had sold at auction, with the Wolfson Foundation paying £100,000 and the state the rest. Reuters then received a note demanding that the £140,000 go towards television licences for impoverished people, which would then ensure the return of the duke, provided that there was an amnesty for the thief. Spike Milligan got involved, saying he supported the demands, but nothing happened further. In 1963, the thief supplied a note from the back of the painting to prove it was still in his possession and then a fourth note stated "I am offering three pennyworth of old Spanish firewood, in exchange for £140,000 of human happiness". The last note in 1965 suggested an exhibition of the painting with paid admission and all the monies going to charity.

'The Daily Mirror' liked the idea and so in May 1965, the painting was returned anonymously to the paper via a left luggage locker at Birmingham New Street railway station. Two months later, a man with the wonderful name of Kempton Bunton walked into a London copshop and confessed to the heist. The police remained doubtful however, since Mr Bunton was a portly old pensioner and it seemed unlikely he could have shinned up a drainpipe, crawled through a toilet window, stolen the painting and then retraced his steps. Bunton stuck to his story and said he did it after becoming enraged that the state spent so much on a mere painting when people were too poor to pay for a TV licence. He had in fact previously served time for not paying his fee. At what was no doubt an entertaining court-case, his barrister Jeremy Hutchinson QC argued that since Bunton had never planned to keep the painting, he could not be convicted of stealing it. This ruse worked, yet Bunton was sentenced to three months in prison on account of the frame being stolen (and not returned). The Theft Act was subsequently altered so that this particular dodge could not be used in court again. A final twist came in 2012, long after Bunton's death, when it was revealed that his son had actually confessed to the crime but had never been charged. He said his dad had sworn him to secrecy! I look forward very much to the forthcoming film adaptation which stars Jim Broadbent as Kempton and Helen Mirren as his long-suffering wife.

Continuing with the duke theme, the 'Buccleuch Madonna' is one rather funny-looking version of Leonardo da Vinci's 'Madonna of the Yardwinder' which ended up hanging in Drumlanrig Castle in Drumfries and Galloway, Scotland. This was until it was stolen in 2003 by two likely lads who told a startled couple from New Zealand "Don't worry love, we're the police. This is just practice" as they left the castle through a window. There was not a sniff of the painting until 2007, when the owner, the Duke of Buccleuch, was contacted by a lawyer offering to broker the return of the painting. At a meeting the lawyer spoke to two people claiming to be representatives of the duke, who were actually cops. They set up a second meeting in Glasgow and the lawyer turned up with the painting, only to be arrested along with four other people. It's unclear why the lawyer believed he wouldn't get arrested, on the other hand the police managed to screw up the case against the arrestees and in the end nobody was convicted. Bizarrely, the lawyer then sued the duke and the police for £4.25 million for breach of contract! To end on a happier note, the painting now hangs in the National Gallery of Scotland, on loan from the Duke of Buccleuch.

Back in the late 18th century, Thomas Gainsborough painted a portrait of Georgiana Cavendish, a social dilettante who was also the Duchess of Devonshire. She sports a magnificent black hat, which apparently started a fashion trend. The painting had a chequered history, disappearing from Chatsworth House (the traditional seat of the dukes of Devonshire despite being up north in Derbyshire) only to be sold at auction in for 10,000 guineas, a shocking amount of money at the time. More shocks followed swiftly afterwards when the painting was stolen from a gallery where it had been on display. No news followed (although there plenty of rumours) until it emerged decades later that the thief had been none other than Adam Worth.

He was a notorious criminal thought to have inspired Moriarty, the nemesis of Sherlock Holmes. Starting as a pickpocket in New York, Worth went on to rob banks and pawnshops in England. He then moved to Paris in 1871, just after the event of the Commune. Amidst all the chaos, he set up a gambling den. Having fled to London, he kept up various criminal enterprises and in 1876 stole the painting with two associates. Apparently Worth took the painting to pay his brother's bail in an unrelated case and when that was no longer necessary he kept the painting for himself. Or he just liked having it around. In any case he only offered to return it in 1901, negotiating a fee of $25,000 for its safe return. He must have needed the money; he died a pauper in 1902 and was buried at Highgate Cemetery. The painting was eventually returned to Chatsworth House by the Duke of Devonshire, but only after he shelled out over £400,000 for it in 1996.

The 'View of Auvers-sur-Oise' was painted by Cézanne in the late 19th century and marks a transition between his early and later works. It is acknowledged as a classic and is estimated to be worth at least £5 million on contemporary rates.

On New Year's Eve 1980, as fireworks exploded and revellers in Oxford celebrated, two thieves scrambled up scaffolding and crept over the roof of the Ashmolean Museum. They smashed a skylight and threw a smoke bomb inside, then descended using a rope ladder. Once on the ground, they used a fan to find their way through the smoke straight up to the Cézanne picture. They grabbed it and smashed it on the floor to remove it from the frame, then rolled up the painting and made their escape, ignoring all the other artworks. Moments later the guards arrived responding to the alarm and because of all the smoke they at first assumed it was an alarm for a fire. It was only later they noticed the tell-tale gap on the wall and the broken glass on the floor. By then, the cheeky thieves were well away.

The police assume that the picture was stolen to order, since it would be impossible to sell on the open market and the thieves took nothing else from the gallery. The museum was forced to admit that the painting had not been insured. It has not been seen since; the police have only had one tip off, which came about a month later when the picture being seen in a pub in the West Midlands. When the cops kicked the doors in, they found a contented amateur painter putting the last touches to their tribute copy of 'View of Auvers-sur-Oise'! Well, it had been in the news.

The Whitworth Gallery in Manchester was subject to a daring heist in April 2013, in which paintings of an estimated value of £4 million were stolen and then quickly returned. Van Gogh's 'Fortifications of Paris with Houses', Gauguin's 'Tahitian Landscape' and Picasso's 'Poverty' were all whisked away from the museum only to appear again the next day in a cardboard tube hidden in a nearby public toilet. A note criticised the woeful security of the museum, although the proud perpetrator probably wants to forget about the six inch rip they made in the Van Gogh. The finger of blame pointed to an ex-member of staff with a grudge yet no-one has ever been caught.

Baronet Henry Price earned his fortune supplying cheap suits to the masses. After he sold his business at massive profit, he bought Wakehurst Place and then donated it to the nation when he died (it's now the Millennium Seed Bank run by Kew Gardens). He drowned whilst sailing to Jamaica in 1963 and his wife Lady Annie Price (nee Craggs) retired to an old rectory in Newick near Haywards Heath in Sussex, only to be disturbed from her blissful rural retirement in the 1980s by a series of burglaries targeting her extensive art collection. In 1984, a number of paintings and valuable objects (such as a clock, a magnifying glass set, a table and four porcelain vases) were stolen, including the 'Portrait of Miss Mathew, later Lady Elizabeth, siting with her dog before a landscape' by Joshua Reynolds.

It's a relatively charming portrait, valued at £1 million by the Telegraph in 2019, which has had a rather unfortunate history. Also known as 'A Young girl and her dog', it was painted around 1780 by Reynolds and his studio. It was often on display in various galleries in the 19th century, then it was sold to the Viscount of Fareham in 1913 and bought in turn from him by Henry Price in 1942 for the sum of £6,000. After the theft, the portrait completely disappeared from sight until 1988, when it was sold at Sothebys in London. It wasn't just sold, it even featured on the cover of the catalogue! Four years after being stolen from a place merely 50 miles to the south, the portrait was bought by a dealer who Sothebys refuses to name. This scumbag, who must have known it was stolen, then sold the artwork on to a Japanese museum, no doubt at a hefty profit. The Tokyo Fuji Art Museum (TFAM) has so far refused to recognise the claim of the Price descendants on their picture.

The whole debacle throws up some interesting points. Obviously first off, it's staggering that between Sothebys, the dealer and TFAM nobody did their due diligence in order to check the provenance of the painting as they are all required to do. The dodgy dealer is one issue but the other two entities are public bodies and supposed to follow the guidelines in such matters. The last sale recorded was the 1913 one, which does beg the question what had happened to the painting in the meantime before it appeared at auction. Under English common

Pretty pictures

law, the 'Portrait of Miss Mathew, later Lady Elizabeth, sitting with her dog before a landscape' still belongs to the heirs of the owners since it was acquired by virtue of theft. However, under Japanese civil law, an entity which acquires an item through legal means is the de facto owner, and should only return the item to the original owner (once jiggery pokery has been exposed) after the amount the entity paid has been recompensed by the original owner. What a mess!

Still though, looked at from a different angle, it seems a rather open and shut case, since the family can provide papers to back the 1942 purchase and the picture was on the front-page of the 'Mid Sussex Times' which announced the loss in 1984. However, the TFAM has form for this sort of thing, having become embroiled in a scandal over a Leonardo da Vinci painting which whilst acquired legally, turned out to have been expropriated from Italy at the end of World War II. The museum tried at first to claim that the Prices had not proved the 'Portrait of Miss Mathew, later Lady Elizabeth, sitting with her dog before a landscape' was indeed the same picture as the work they know as 'A young girl and her dog' and is now saying it will return the painting, but only if it gets compensated in full. This is all far too late for Annie, since she died in 1994.

It's not just paintings that get nicked of course, anyone who has watched 'Fleabag' knows that sculptures can also go walkabout. Sadly with large works, their scrap-value can be more attractive to some than their value as art. Henry Moore made a series of sculptures under the name 'Reclining Figure 1969-1970' and in 2005, the artist's cast was stolen from the Henry Moore Foundation in Hertfordshire. The ingenious barbarians winched it onto a flatbed truck and drove the two tonnes of newly acquired scrap-metal away, never to be seen again. An art piece estimated to be worth £3 million was probably sold off for around £1,500. Now that's serious art criticism! A bronze sculpture by Barbara Hepworth was stolen from Dulwich Park in similar circumstances in 2011.

Last but by no means least, we have the entertaining story of 'America'. This was the name of a functioning golden toilet made by the Italian artist Maurizio Cattelan. He produced it in 2016 to be installed alongside the regular toilets at the Guggenheim Museum in New York. The gallery made the genius publicity step in 2017 of offering it to the newly elected Donald Trump as a gift and proudly reported over 100,000 people had queued to use it (maybe even take a shit). Trump meanwhile did not respond (for once).

In 2019, the participatory (pissapatory?) sculpture arrived at Blenheim Palace, where it was plumbed in and exhibited. Then of course it was stolen, or else it wouldn't be mentioned here. "Busted flush" screamed 'The Guardian'. Pressed for a comment, Cattelan said "I always liked heist movies and now I'm in one of them".

He himself is no stranger to theft, having stolen the entire contents of a Dutch art gallery and exhibited it as art in the 1990s. What goes around comes around.

Six people were arrested as police investigated the theft of the toilet and a reward of £100,000 was offered, but privately officers are convinced the toilet would have been melted down as soon as possible.

Pretty pictures

CHAPTER SEVEN

On a cold Sunday evening in December 2004, two police officers knocked on the door of Kevin McMullan's converted farmhouse in Poleglass, a village to the west of Belfast. McMullan had spent the weekend relaxing and came to the door with a smile. The officers said they need to report that a family member had come to harm in a traffic accident. Then, as soon as they were inside the door, McMullan and his wife Karyn were looking down the barrel of a gun. They were taken hostage and threatened with their lives. So began the Northern Bank robbery, setting off a chain of events which would eventually almost derail the Northern Irish peace process.

Whilst nobody has been held directly responsible, there have been many arrests. For some people such as the weasel-faced Tony Blair and Bertie Ahern, the Irish Taoiseach, it was clear that the IRA was behind it, a charge hotly denied by trustworthy truth-tellers such as Gerry Adams and Martin McGuinness. Over 15 years later the story is still not over and related controversies continue to reverberate.

Whilst the McMullans trembled in fear, Chris Ward was coming to the door of his house in Downpatrick, a small town southeast of Belfast.

A man was knocking to ask about tickets for the next Celtic football club match, a not uncommon event since Ward was involved in the fan-club. Suddenly however, a gun appeared and Ward and his family were threatened with their lives. In contrast to Karyn McMullan who was hooded, forced to put on a boiler suit and dragged off to an undisclosed location, the armed men took Ward with them leaving the hostages behind with two guards. Later, this difference in treatment by the gangsters would come back to bite Ward in the arse. He was ordered to lie down in the back of a car, then driven to Poleglass. The criminals changed vehicles on the way.

Ward worked at the headquarters of the Northern Bank in central Belfast, as did McMullan, who was his supervisor. It was an impressive building which stood on the same central square as the City Hall and the Linen Hall Library. The two men had access codes which got them into the vaults beneath the bank where the cash was stored and since Northern Bank could print and dispose of its own money, there was always a lot of cash passing through the headquarters. Once they had the two officials at their beck and call, the perpetrators would not even need to enter the bank

The heist that broke the peace process

themselves to execute their fiendish plan. The gang left the Poleglass house early in the morning of Monday 20 December 2004 and told Ward and McMullan to follow their instructions to the letter. They were to go to work just like normal, with new mobiles in their pockets they had to keep on at all times. Ward's family were left with two men who told them if they didn't do anything stupid, no-one would get hurt. The bank workers travelled to the Belfast offices and put in a normal shift. They must have been scared shitless, but nobody twigged anything untoward was happening. In the afternoon, Ward got a message to fill up a bag with £1 million in high denomination used banknotes and to take it to a bus stop on Queen's Street. When he got there, a man leaning against the stand thanked Ward for the early Christmas present and quickly disappeared.

This had been a test run to see if the bank officials could be trusted. The lives of their loved ones depended upon their actions, so Ward and McMullan stayed fully compliant. They stayed at work late until everyone else had left and they told the security guards that a van was coming to take away office trash. They filled up crates with bags of used bank notes and wheeled them on trolleys covered with broken office furniture down to the loading bay of the bank, where a white Ford Transit van was waiting. Gang members dressed as delivery men helped them to load the van then sped off. An hour later they were back and the van was filled up again. By 21.00 it was all over.

At 23.00, Karyn McMullan was driven to the Drumkeeragh Forest park and thrown out onto the ground. She struggled to her feet and was nearly knocked over by another car. She wandered about for a while before finding a place to raise the alarm. She was treated for hypothermia but she was at least still in one piece. Fortunately, Ward's family were also unharmed, although for them the ordeal was just beginning.

The gang must have been celebrating, since they had got away scot-free with over £25 million in cash. When the newspapers heard about it, the heist was quickly declared the biggest in UK history although there are several other entries in this book which would like to have a word about that. What we can say with certainty is that it was Northern Ireland's largest bank robbery. The criminals made off with a grand total of £26.5 million. In fact, Ward later said they got away with much more! The total that

The heist that broke the peace process

was announced to the press was comprised of a cool £10 million in newly produced Northern Bank pound banknotes; there was another £10 million in used banknotes which had been in circulation; and making up the other £6.5 million were stashes of cash in other currencies such as US dollars and euros. Not a bad haul at all!

Chief Constable of Northern Ireland Hugh Orde tried to play down the extent of the heist by saying it was the "the biggest theft of waste paper in history". Whilst it is true that months later Northern Bank released new banknotes, meaning that the fresh stolen notes were invalidated, that still left at least £16.5 million of untraceable cash! Orde was correct however when he said that the heist bore all the hallmarks of an operation run by the Provisional Irish Republican Army (better known as the IRA), since it was meticulously well-planned and audacious in its scope. Just to rub Loyalist noses in it, another unnamed senior cop said groups like the Ulster Volunteer Force (UVF) and the Ulster Defence Association (UDA) did not have the capacity to do it. In fact, the robbery followed several others in previous months which had a similar template of both hostage taking and military organisation. In

October, the Gallaher Group cigarette warehouse in Belfast had been looted to the tune of £2 million, at Easter £1 million of goods was taken from a Makro store and in summer an Ulster Bank on the border with the Republic had been robbed of £500,000. The authorities were tight-lipped about who had done the heists, but they would go so far as admitting they suspected they had all been carried out by paramilitaries. Perhaps other armed sectarian groups were also getting things done whilst they still could but from its scale, it was obvious to most commentators that the IRA was behind the Northern Bank robbery. Reasons were easy to find for why the heist had happened precisely then, since the peace process was grinding on and pretty soon the IRA would be required to decommission its weaponry and cease its illegal activities. If the IRA did have to demilitarise, then a large robbery would fill the coffers for a pension fund for its recruits and perhaps even fund a political campaign. The more cynical view was that the IRA was simply carrying out all its long-held plans while it still had the opportunity; the funds could even pay for a covert plan to buy more weapons.

An answer came directly from the IRA in mid January when it released a terse statement: "The

IRA has been accused of involvement in the recent Northern Bank robbery. We were not involved." Not many people believed it. The roadmap for the peace process was clear even if not many people could see it actually happening, despite the rhetoric of coves like Gerry Adams. Adams was a Sinn Féin representative, Sinn Féin being the political mouthpiece of the IRA. Even to this day he denies that he had ever been in the IRA, despite evidence to the contrary, so much so that it is said that he has a beard purely so he can cover up all his bare-faced lies. In fact, more than being a member of the IRA, Adams is commonly known by analysts of the Troubles (such as Richard English and Ed Moloney) to have been a member of the IRA Army Council from 1977 until as late as 2005. Kevin Myers even states in his book 'Watching the Door: Drinking up, getting down and cheating death in 1970s Belfast' that he once heard Adams casually give the order to execute a man. Still it doesn't really matter here if Adams was an IRA leader or not, what matters is that as a Sinn Féin leader, he would in any case have been aware of a plan of this magnitude, despite his denials.

Having been caught with their pants down the Police Service of Northern Ireland (PSNI) naturally threw everything they had at the case, but as a decades old clandestine organisation the IRA did already have substantial and long-reaching networks set up to wash the money and the PSNI was hampered by its Special Branch having been wound down as part of the peace process. Further, committed paramilitaries waging a decades long war against an invading state get quite good at minimising their forensics, so the cops didn't have much at all to go on at the crime scenes. On the other hand, twenty million quid is a lot of money to launder ...

A report was released in February 2005 by the Independent Monitoring Commission (IMC) which blamed both Sinn Féin and the IRA for the heist. The IMC was a supposedly impartial group set up to oversee the police process although radical republicans were sceptical of its position, saying it was led by three spooks and a lord. The IMC suggested fining Sinn Féin and this resulted in a political scandal. With timing denounced as opportunistic, the PSNI launched raids in Beragh, County Tyrone, on the same day as the report was launched. They didn't turn up anything, but if you shake the tree sometimes an apple will fall out.

The heist that broke the peace process

One week later, the Garda Síochána (Ireland's police force), launched an operation against money laundering which turned up a few people of interest. Seven people were arrested and over £2 million in cash was seized, of which at least £60,000 could immediately be linked to the Northern bank robbery. The sixty grand was seized from a car searched at Heuston station in Dublin; it was carried in a washing powder box by Derek Bullman who was later convicted of membership of the IRA, getting four years in the cell. At his trial, Bullman said he was attending a catering exhibition and the court was told along with the cash, a name-tag for Jerry McCabe, Catering Officer, Garda Club, was recovered from the car. Bullman said this was a joke and prosecutors said it was not a funny one, since Detective Garda Jerry McCabe had been killed by the IRA in 1996.

Another person arrested was Ted Cunningham, a financial adviser. His home in Farran, County Cork, was raided and a stash of £2.3 million was found in the most unlikeliest of places, a compost heap in his garden. Neighbours in the sleepy village where he lived expressed surprise at his arrest, with one person giving a classic line to the BBC: "Some thought he might be up to the odd shady deal, but nothing like this".

Others said he liked a singsong down the pub and for the rest, kept himself to himself.

At Cunningham's trial, the police alleged that he had collected almost £5 million in notes from the robbery in four separate loads. At first, Cunningham explained away the money as investments from Bulgarian businessmen wanting to capitalise upon the 1999 switch to the euro in the Republic. The police did not believe him and claimed he had already told them when under interrogation that he had concocted an elaborate alibi featuring the Bulgarians wanting to buy a pit he owned near the village of Shinrone, which would then allow him to launder the money. When they asked him if he knew the cash was from the Northern Bank robbery, he had confessed that as soon as he saw the Northern Irish banknotes he knew it was from the heist. Cunningham and his son were both convicted.

The controversy only deepened when the well-known banker Phil Flynn announced he had visited the police on his own initiative since he held a 10% stake as a non-executive director in Chesterton Finance, a company set up by none other than Cunningham. Flynn was a big cheese: he was chairman of the Bank of

Scotland in Ireland and he had the ear of the Taoiseach. The police said Cunningham was working for Phil Flynn in order to launder the loot. Cunningham argued that he was licensed to hold large amounts of cash as a financier, yet things only worsened for him when a Cork man want to the police and gave them £175,000 which he said he had been given by Cunningham.

The political scandal which was brewing was huge. To provide a very brief overview, the Good Friday Agreement of 1998 had brought the beginnings of an end to the Irish Troubles. With pressure from the USA and the governments of the Republic of Ireland and the United Kingdom, plus behind the scenes networking of many other power brokers including community leaders and the Catholic church, things were heading towards a settlement in which power would be shared in a devolved Northern Irish government at Stormont. Various bumps in the process, such as the discovery of a British spy ring, were already threatening to derail the fragile peace, but the twin facts both that the heist had even occurred and that responsibility for it was denied by the IRA really irked the other participants, who had been pressing the IRA to drop its illegal activities and come to the negotiation table.

Throughout 2004, in a series of top level meetings, the Irish and British governments had pressed Sinn Féin to compel the IRA to cease its illegal activities as an important and necessary step in the peace process. Obviously, Sinn Féin was quite unable to persuade the IRA to do this when a massive heist was in the works. At a tense meeting in Belfast two weeks before the Northern Bank robbery took place, Gerry Adams and his righthand man Martin McGuinness had again refused to make any promises. This led to disgusted reactions from the other participants.

Bertie Ahern was at the time the Taoiseach, the prime minister of the Republic of Ireland. He was pretty forthright in his views, saying "it is of concern to me that an operation of this magnitude was obviously being planned at a stage when I was in negotiations with those who would know the leadership of the provisional movement and that raises questions, and questions that concern me".

This drew a quick response from Adams, who commented "now the Taoiseach really believes, and he made the remark, and he told me he made it in a considered way then he has no option but to have us arrested, to put up or shut

The heist that broke the peace process

up on this issue." 'The Irish Independent' commented "there were amused and knowing giggles from political and security sources in Belfast, London and Dublin yesterday about Mr Adams's claims to know nothing about the Northern Bank robbery." Adams has always carried a stench and it is instructive to wonder if he himself is a British agent, since otherwise it's hard to understand how he hasn't been killed or locked up. Ahern's ire was probably based on his knowledge that Adams had been observed by Garda surveillance meeting up with Cunningham in the weeks before the heist.

And so let's return again to Ted Cunningham. He had been convicted of money laundering in 2009 but on appeal to the Court of Criminal Appeal, his conviction was set aside and a retrial ordered. At the retrial in 2014 (ten years after the heist!) Cunningham was again convicted but escaped imprisonment on account of time served and sympathy for a medical condition that had apparently led to him collapsing in jail 16 times. He pled guilty to two charges and received a suspended sentence. But that was not the end of it, far from it, since the Cunningham saga extends into the present. Just when you thought it was all over, Cunningham launched a judicial review application in 2018 regarding the way in which the money confiscated from him had been handled. He named as plaintiffs the Garda Commissioner, the Head of the Criminal Assets Bureau, the Director of Public Prosecutions and Danske Bank (the new owners of Northern Bank). The review was refused; he then appealed the decision and was refused again, when he ran out of time to make the application. But the story still keeps going! In 2020, fully fifteen years after his arrest, Cunningham was again going back to court! This time he was attempting to argue that the money had been illegally impounded by the police. You can't blame him for trying but you have to wonder why he doesn't give up. Is he innocent? I don't think so.

Turning back the clock to 2005, whilst the arrests of Cunningham and others in the Republic were generating news, an amusing side-story occurred when the PSNI was forced to confirm that five neatly shrink-wrapped packets each containing £10,000 had been discovered in the toilets of the Newforge social club in Belfast after an anonymous tipoff. Newforge was a private members club exclusively for cops and former cops. Like the still unexplained robbery at Castlereagh police station in 2002, this prank shows the amazing

The heist that broke the peace process

power and reach of the IRA. Whilst the station raid could also have been Mi5 covering its tracks (and is worthy of its own chapter in another book), the dumping of the cash was definitely IRA and a cheeky riposte to the arrests down south. When it is born in mind that the bundles were probably composed of the notes which would have been easy to track if spent, it's a cheeky action all round!

Shaking of various trees occurred later in 2005, when more people were arrested, but nothing linked back directly to the Northern Bank robbery. Then in a startling new plot twist worthy of the finest soap opera, Chris Ward was arrested in November 2005. According to the police, their suspicions were first raised by his behaviour during the heist. Remember that whilst Karyn McMullan was blindfolded and taken elsewhere, Ward's family was kept at his house and that Ward had been the one to take a bag of cash out as the test-run. However, it quickly emerged that there was no case to answer: the main plank of the prosecution's argument was that Ward had manipulated his work rota so that he was working on the day of the heist. Ward himself complained bitterly at trial that he was being prosecuted for being Catholic and that the police had been bugging him and his family for months (indeed, the prosecution wanted to use tapes recorded whilst Ward was on holiday in Spain). The social club where Ward worked was raided but nothing was found. Was Ward really guilty? The case dragged on and on, with a trial eventually being set for September 2008, almost three years after Ward's arrest. What a horrible time that must have been for him and his loved ones. When the date of the trial eventually came, the prosecution immediately folded, saying that it now admitted that it was Ward's bosses who had changed the work rota and not him, so without that flimsy part the whole edifice of the prosecution collapsed. What a joke, I hope he sued!

The scandal of the unsolved heist continues to rumble on into the present. In 2010, Bertie Ahern repeated his view that the IRA was responsible, backed by both Dermot Ahern (no relation) who had been Minister for Foreign Affairs in 2004 and Mary Harney who had been the Tánaiste, the deputy head of the government of Ireland. In 2014, the former head of the Assets Recovery Agency commented that he believed the IRA were still struggling to launder some of the £26 million taken in the heist owing to the size of the haul. We are now in 2020 and still nobody has been held directly responsible, although the fallout continues. As various IRA legends die, the Northern Bank robbery crops up again. For example, big Bobby Storey finally went to meet the angels in the summer of 2020 and once expired was named as the architect of the robbery.

Ted Cunningham is of course still trying to prove his innocence and announced in 2020 he was launching a sweeping lawsuit against those he claimed had falsely imprisoned him. Amongst others, he planned to sue Justice Minister Charlie Flanagan, Garda Commissioner Drew Harris, Director of Public Prosecutions Claire Loftus and Attorney General, Seamus Woulfe. He said "What happened to me was wrong — the State just decided that the money that they took from me was from the Northern Bank and that was it."

Surprisingly, the events of the Northern Bank robbery have not been made into a Hollywood film (yet). Perhaps when a few more participants have died off the truth will come out and we'll get a thriller with Robert Redford as Bobby Storey, a ferret as Tony Blair, Timothy Spall as Hugh Orde, and Pierce Brosnan as Gerry Adams (he's born to play him, in fact it already happened in the crappy Jackie Chan vehicle 'The Foreigner').

We do have books and plays though! Before this little book stormed the shelves, former IRA bank robber Ricky O'Rawe published 'Northern Heist' about a Belfast bank heist which he claimed was fiction and a play was performed in Poleglass in 2010 which aimed to focus on the stories of those people who were taken hostage.

CHAPTER EIGHT

To conclude this little book, I'd like to revisit some of the most entertaining heists that we know about. Of course, there may be many other ones we don't know about ...

In 2020, the Cambridge University Library announced rather sheepishly that they had not been able to find two of Charles Darwin's notebooks (deemed to be priceless) since 2000 and that they were starting to think they might have been stolen. Apparently the notebooks were taken to be photographed in November 2000 and then nobody thought to look for them again until January 2001, when they were discovered to be missing. The library does have 200 kilometres of shelving and sure I also lose things down the back of the sofa occasionally, but it seems rather lax to simply say "oh they'll turn up" and then wait for twenty YEARS. Well in any case by 2020 the library had still not found them and it will take another five years to do a full search.

So it decided to bite the bullet and admit that the notebooks may well have been nicked ... twenty years ago already!

You either laugh or you cry

It turns out the British Library has also misplaced items. It admitted in 2017 that it was unable to find a £750,000 Cartier ring and it had been lost since 2011.

And it gets worse! Sky News ran a disturbing story in 2017 stating that since 2010, a total of 947 items had gone missing from British museums ...

Now I'd like to end with one of my fave stories.

It involves workers at the Bank of England's printing and incineration works in Essex stealing money and it's a real David and Goliath story! Between 1999 and 1992, a syndicate of four workers is believed to have stolen around £600,000 in used notes which had been taken out of circulation and were waiting to be destroyed. In a way, it was a victimless crime since the notes were trash, although it has to be said that the Bank of England didn't see it that way. The ringleader was Christine Gibson, who became an instant tabloid heroine when it was revealed she had smuggled notes home in her underwear. Gibson and her husband enjoyed seeing the world. They holidayed in Hawaii, the

Bahamas, the Nile and the Far East, despite her job only paying her around £16,000 a year. At the time of their arrests, they had two cars, two motorbikes, a horse, lots of of jewellery and a mortgage free house with £600 in her underwear drawer!

They were caught only because the husband in his great wisdom went to a local building society and tried to deposit £100,000 in twenty and fifty pound notes, which triggered an automatic drug-dealing investigation. The staff's suspicions were aroused further when an accomplice also attempted to deposit £30,000 in used notes, claiming to have found them in their parents' home. Things unravelled totally when another worker named Kevin Winwright was arrested and confessed to stealing £170,000. He turned grass and informed on the rest, yet strangely he was convicted and they were not; he served 18 months and the others stayed free, mostly because they refused to admit to anything and back in those days before unexplained wealth orders, an extravagant lifestyle was not in itself enough to get a conviction.

However, the Bank of England then began a civil case to recover the money and the judge had their number, describing their flannel as "wholly incredible" and commenting "the defendants say that unless I can make specific findings that such-and-such was stolen by so-and-so on a particular date on such a sum, I am not in a position to make any findings against the defendants at all. So far as that is concerned, I reject that submission."

In fact the judge seems like a geezer, he also said: "Mr Gibson told me that he is a very hard working man, now being some 47 years old. The bank's case is that he has never done an honest day's work in his life. His answer to that is "Not at all. I have never passed a day of my life effectively when I have not worked." The honesty may be a matter of debate; nevertheless, work he has always done. He says that he always works for cash. He is a great believer in cash, and he went as far as to say he has a genetic disposition to it. "It's in my genes", he said."

Despite all their wiles, the Gibsons were ordered to return £250,000 to the Bank of England and to pay their share of the court costs, which probably reached £250,000 as well. Regarding the horse, the judge could only say: "I know little except that it is now dead and it was a Cleveland Bay."

Ultimately, the judge had some fun and people had to return some money. However, it was never identified just how much had been stolen in the first place and nobody went to jail except the snitch. Naturally, two movies were made about these shenanigans, firstly an ITV film called 'Hot Money' starring Caroline Quentin and then the inevitable Hollywood version entitled 'Mad Money', which starred Diane Keaton, Queen Latifah and ... Katie Holmes!?

To wrap this up, I'll give a final quote from the judge, who should probably have got his own standup comedy show: "I do bear in mind, as Mr Wadsworth asked me to do, that I live in the real world and that I should not be surprised at people earning money for cash. I do live in the real world and I am not surprised that people earn money for cash. It is probably a very regular occurrence over a large part of the population. But £30,000 to buy bonds, £100,000 in cash to buy bonds, this level of income is in my view wholly exceptional and simply not believable."

You either laugh or you cry

PEBBLE POCKET BOOKS

ONE
E. T. C. Dee *Harden Noten: Voices from the Rotterdam squatters movement*

TWO
T. Stolinski *Rockbottom Rotterdam*

THREE
Kitty Lyons *The Little Book of UK Heists*

FOUR
T. Stolinski *Popcorn*

FIVE
Using Space

Ingram Content Group UK Ltd.
Milton Keynes UK
UKHW020013020523
421049UK00016B/1052